Wackiest Machines Ever!

Paul Mason

www.raintreepublishers.co.uk
Visit our website to find out more information about **Raintree** books.

To order:
☎ Phone 44 (0) 1865 888112
▤ Send a fax to 44 (0) 1865 314091
💻 Visit the Raintree bookshop at **www.raintreepublishers.co.uk** to browse our catalogue and order online.

First published in Great Britain by Raintree, Halley Court, Jordan Hill, Oxford OX2 8EJ, part of Harcourt Education.
Raintree is a registered trademark of Harcourt Education Ltd.

© Harcourt Education Ltd 2006
First published in paperback in 2007
The moral right of the proprietor has been asserted.

Editorial: Lucy Thunder, Charlotte Guillain, Richard Woodham, and Harriet Milles
Design: Michelle Lisseter, Carolyn Gibson and Bigtop
Illustrations: Darren Lingard
Picture Research: Melissa Allison and Virginia Stroud-Lewis
Production: Camilla Crask

Originated by Dot Gradations Ltd.
Printed and bound in Italy by Printer Trento srl

The paper used to print this book comes from sustainable resources.

ISBN 1 844 43844 9 (hardback)
10 09 08 07 06
10 9 8 7 6 5 4 3 2 1

ISBN 1 844 43859 7 (paperback)
11 10 09 08 07
10 9 8 7 6 5 4 3 2 1

British Library Cataloguing in Publication Data
Mason, Paul
Wackiest Machines Ever!: Energy– (Fusion)
531.6
A full catalogue record for this book is available from the British Library.

Acknowledgements
The publishers would like to thank the following for permission to reproduce photographs: Corbis/Bettmann pp. 22–23; Corbis (Chris Hellier) p. 8; Corbis/Hulton-Deutsch Collection pp. 4–5; Corbis/Reuters p. 25; Empics/PA pp. 6–7; Motoring Picture Library pp. 18–19; Reuters (David Hancock) pp. 17, 26–27; Rex Features p. 14; Science and Society Picture Library/Science Museum pp. 20–21; The Kobal Collection/DANJAQ/EON/UA pp. 12–13.

Cover photograph of a rocketman, reproduced with permission of Alamy (James Callaghan).

Every effort has been made to contact copyright holders of any material reproduced in this book. Any omissions will be rectified in subsequent printings if notice is given to the publishers.

The publishers would like to thank Nancy Harris and Harold Pratt for their assistance in the preparation of this book.

Contents

Some words are printed in bold, **like this**. You can find out what they mean on page 30. You can also look in the box at the bottom of the page where they first appear.

How wacky does it get?

Welcome to the Museum of Wacky Machines. Some of these crazy machines actually moved. Others never got off the ground.

All the machines in this book need **energy** to make them move. Energy is the ability to make a change happen. You need energy to power a flying machine. You need energy to power a record-breaking car. You also need energy when you run for the bus!

energy   ability to move something or make something happen

The people who made these wacky machines found some wacky ways to make them move. Read on to find out more!

▼ *This crazy-looking machine is an ornithopter. You can find out more about it on page 20.*

Forms of energy

It is helpful to think of **energy** in two forms. The first form is called **potential energy**. The second form is called **kinetic energy**.

This man is taking part in ▶ a Birdman competition. In these competitions, people try to fly using wacky machines.

kinetic energy   energy of movement
potential energy   stored, or possible, energy

The birdman uses both forms of energy at different times. At first he is just standing on the pier. He is not moving. But he may possibly jump from the pier. This possibility for movement is called potential energy.

When the birdman steps off the pier he starts moving downwards. This is kinetic energy. It is the energy of movement. Kinetic energy always needs some sort of movement.

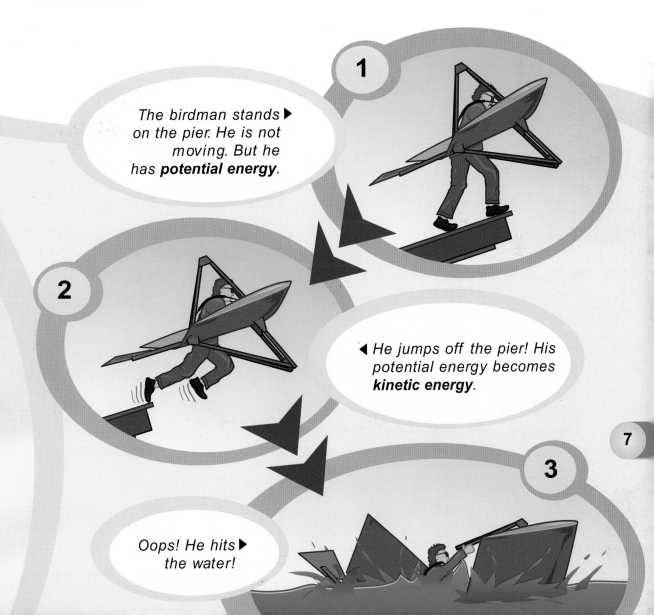

1

*The birdman stands ▶ on the pier. He is not moving. But he has **potential energy**.*

2

*◀ He jumps off the pier! His potential energy becomes **kinetic energy**.*

3

Oops! He hits ▶ the water!

The trebuchet

The trebuchet was used hundreds of years ago. It was a very powerful weapon. It could even break down the walls of a castle.

The trebuchet had a giant arm. It worked like a see-saw. At the short end of the arm was a heavy weight. At the long end of the arm was a basket, or cradle. The cradle held a heavy stone. The stone was thrown by the trebuchet.

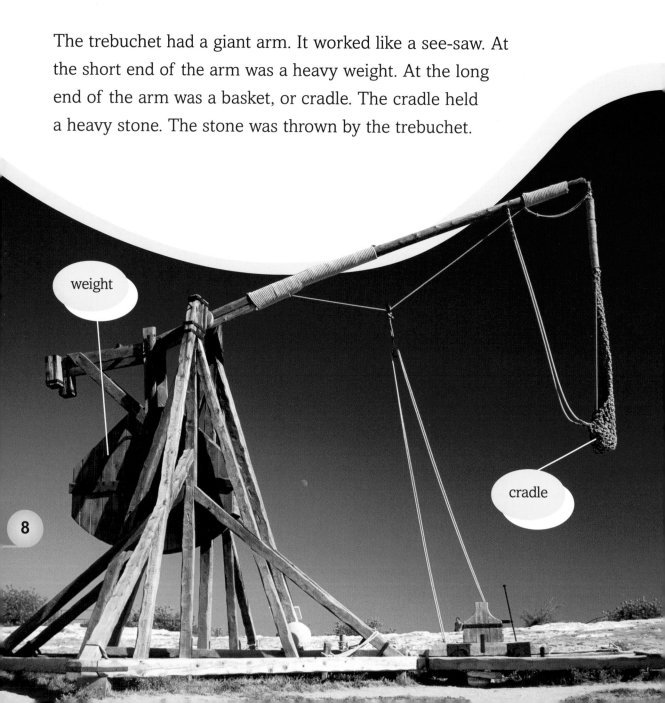

weight

cradle

Soldiers used active, **kinetic energy** to raise the weight forwards and upwards. Then the weight was fixed in place. This meant that the kinetic energy could be stored as **potential energy**. The trebuchet's potential energy was stored until the soldiers were ready to attack!

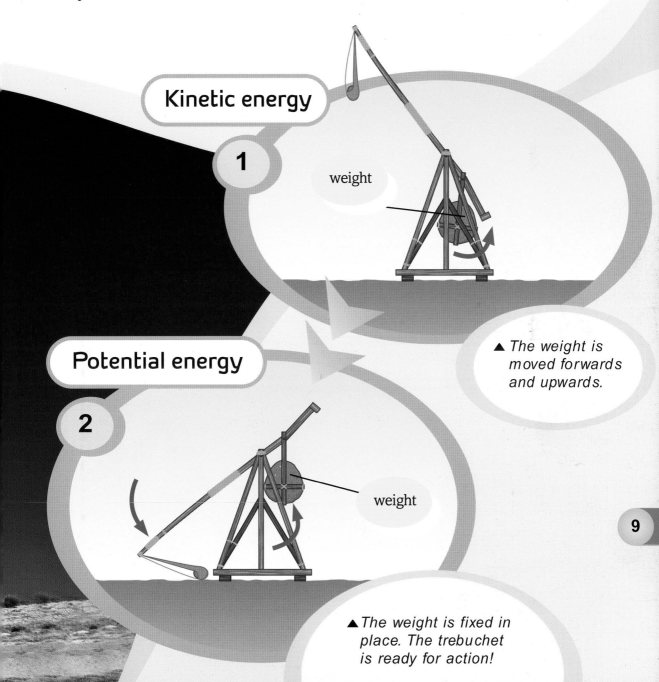

Kinetic energy

1

weight

▲ The weight is moved forwards and upwards.

Potential energy

2

weight

▲ The weight is fixed in place. The trebuchet is ready for action!

9

Attack!

The trebuchet's weight is pulled up and fixed in place. The machine is loaded with stored, **potential energy**. This means it is ready to be moved for an attack.

The attack begins! Soldiers release the trebuchet's weight. The weight drops downwards. The machine's potential energy has become moving, **kinetic energy**. The long arm swings upwards. The giant stone is thrown out of the cradle. It flies through the air towards the enemy!

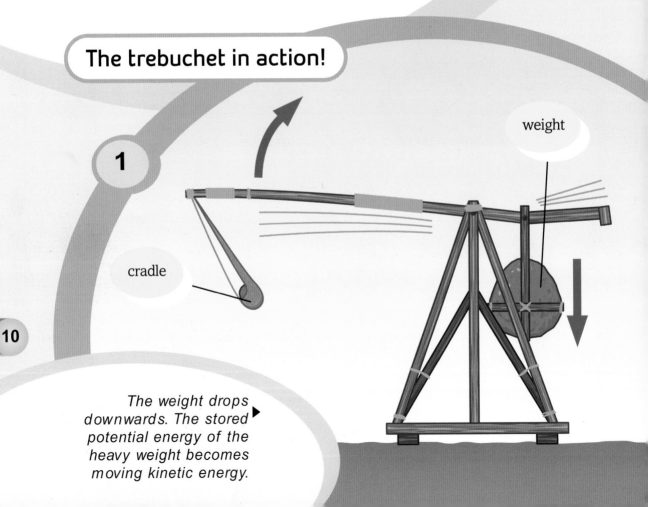

The trebuchet in action!

1

weight

cradle

The weight drops downwards. The stored potential energy of the heavy weight becomes moving kinetic energy. ▶

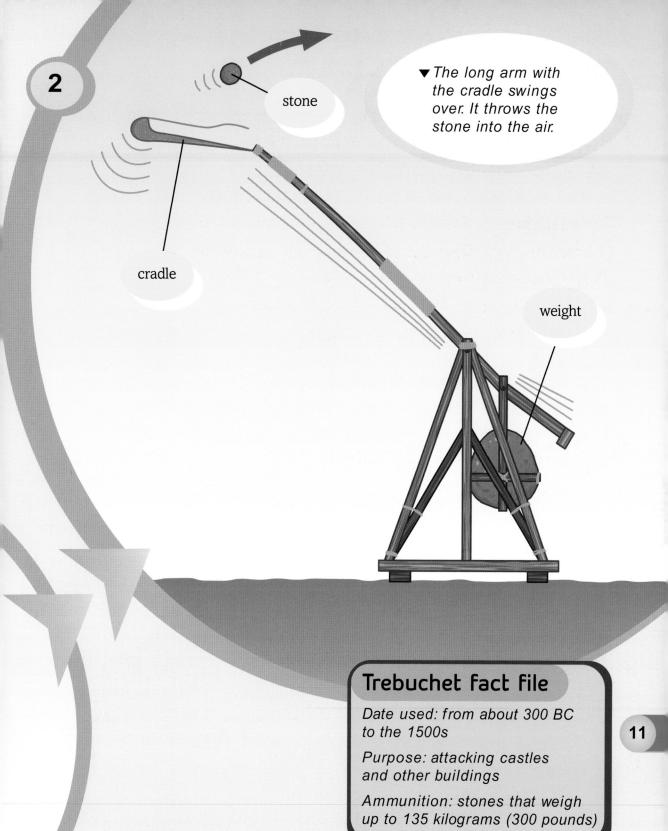

2

stone

cradle

weight

▼ The long arm with the cradle swings over. It throws the stone into the air.

Trebuchet fact file

Date used: from about 300 BC to the 1500s

Purpose: attacking castles and other buildings

Ammunition: stones that weigh up to 135 kilograms (300 pounds)

The rocket pack

The idea of a rocket pack first appeared in a 1920s comic strip. It was worn by a comic-book hero. His name was Buck Rogers. In the 1950s, pilots used real rocket packs.

Rocket packs were worn like backpacks. They lifted the pilots up into the air. But how did the amazing rocket packs work?

The rocket pack's **potential energy** was stored in its **fuel**. The fuel was burned to make hot gases. The burning fuel turned the stored potential energy into the **kinetic energy** of movement. Very hot gases have a lot of **heat energy**.

Rocket pack fact file

First used: 1958
Speed: over 160 kilometres (100 miles) per hour
Fuel used: 1 gallon (4.5 litres) every 5 seconds

fuel   something that produces heat or power when it burns
heat energy   the energy in heat, such as steam or fire

▼ The rocket pack first became famous when it was used in a James Bond film. The rocket pack in the film was not a special effect. It was real!

13

▼The rocket pack did not become popular! This was because it used too much fuel. It used up 6 gallons (27 litres) of fuel in just 30 seconds!

How did the rocket pack work?

The rocket pack's tanks contain **fuel**. Fuel can produce heat, or **thermal energy**, when it burns. When the fuel in the rocket pack burns, it makes heat and hot gases. The hot gases whoosh out of the tubes on the back of the rocket pack. This pushes the pack and the pilot up into the air.

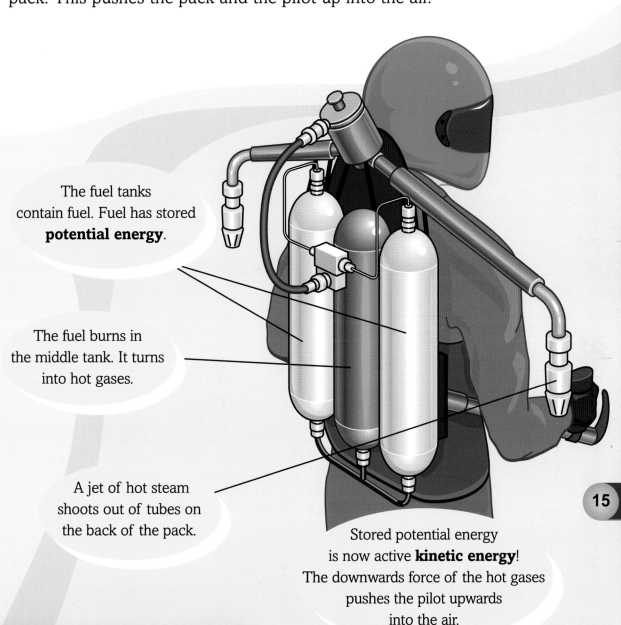

The fuel tanks contain fuel. Fuel has stored **potential energy**.

The fuel burns in the middle tank. It turns into hot gases.

A jet of hot steam shoots out of tubes on the back of the pack.

Stored potential energy is now active **kinetic energy**! The downwards force of the hot gases pushes the pilot upwards into the air.

thermal energy   energy of heat

The steam-powered car

The Shearer was one of the first steam-driven cars. It was made in Australia a long time ago. It used wood and water as **fuel**!

The Shearer car could only move very slowly. Only a small amount of the stored **potential energy** in the car's fuel was turned into **kinetic energy** to move the wheels. The rest of the stored energy was "lost" in other ways.

How machines lose energy

Energy never disappears. It just changes its form. These are some of the forms of **energy** that machines often cannot use:

- **Thermal energy**, or heat energy, can escape into the air.

- The noise that machines make is caused by **vibrations** in the air. These vibrations are a form of energy called **sound energy**.

- When a machine's moving parts rub together, they use up energy.

- When a moving object pushes against air, it uses up energy.

sound energy   energy in the form of noise or sounds
vibration   fast backwards and forwards movement

Shearer car fact file

First driven: 1897
Fuel: wood
Top speed: less than 15 kilometres (10 miles) per hour

1

Fuel (potential energy) is burned in this boiler to produce steam (**heat energy**).

2

Steam pushes a piston, which turns this driveshaft (kinetic energy).

3

Gears below the car turn the wheels.

How fast did the Stanley Steamer run?

The Stanley Steamer was another steam-powered car. It was made in the USA by the Stanley brothers. It ran better than the Shearer.

The Stanley Steamer turned **potential energy** (stored energy) in **fuel** into high-speed movement (**kinetic energy**). At the Daytona speed trials in 1906, it reached 204 kilometres (127 miles) per hour. This was a new speed record.

How did the Stanley brothers make their car run so fast?

- The shape of the car cut through the air easily.
- The boiler could reach very high temperatures. This meant it made a lot of **heat energy** (**thermal energy**).
- Very little steam (heat energy) escaped through gaps in the car's moving parts.

Stanley Steamer fact file

First driven: 1906
Top speed: about 204 kilometres (127 miles) per hour

▼ The Stanley Steamer once ran at 317 kilometres (197 miles) per hour. It was so fast that it crashed before it could complete the course. The speed was never made an official record.

The ornithopter

What is an ornithopter? It is a flying machine with wings. The wings flap like a bird's. Modern aircraft get an upward **lift** from their wings. They get a forward push, or **thrust**, from a propeller or jet engines. The ornithopter tries to get lift and thrust from its flapping wings.

The ornithopter in the photo was powered by a human. Its **energy** came from the pilot. The stored **potential energy** was in the pilot's muscles. It became **kinetic energy** as he flapped the wings to make them move.

The flapping wings were supposed to lift the machine up and drive it forwards. It did not work. The ornithopter did not get off the ground!

Kinetic energy is used to move the wings and flap them up and down.

lift   upwards movement
thrust sudden push forwards, backwards, or upwards

The pilot gets energy from food. The energy is stored in the body as potential energy.

▼ *This is another example of an ornithopter.*

How the ornithopter loses energy

- **Friction** between the moving parts of the ornithopter creates heat. This loses **thermal energy (heat energy)**.
- The ornithopter makes a lot of noise. This loses **sound energy**.
- The pilot gets hot. He loses thermal energy.

LE SAUTERAL Nº1

Why won't the ornithopter fly?

Ornithopters are meant to fly in the same way as birds. But a human body is heavier than a bird's body. So, lifting an ornithopter into the air needs lots of **kinetic energy**. It needs more **energy** than the human body can produce!

There is another problem with the ornithopter. Not all of a human's **potential energy** can power an ornithopter. The human body needs its energy for other things, too. For example:

- The human body needs to use energy to keep the heart, lungs, and brain working.

- When humans take exercise, they get hot. They sweat and breathe heavily. The energy in their bodies becomes heat energy. This energy gets lost in the air.

friction   when two things rub together

Light fantastics!

Think of a car race. The cars in the race travel more than 3000 kilometres (1860 miles) in less than 36 hours. But they do not run on petrol or gas. They run on sunlight!

Energy from the Sun travels to Earth in the form of light. The cars have special **solar panels**. These panels trap the **light energy** from the Sun. They turn sunlight into energy. This energy is used to move the wheels of the cars.

In 2001, a **solar-powered** car called Nuna II won a special race. The race was in Australia. Nuna's average speed was 91 kilometres (57 miles) per hour.

Energy changes

How do solar-powered cars run on sunlight?
- *Sunlight falls on special solar panels.*
- *The panels trap the Sun's energy. They send this energy to a motor.*
- *The motor turns the car's wheels.*
*Sunlight has become active **kinetic energy**!*

light energy	energy in the form of light
solar panel	special panel that traps the Sun's rays
solar power	power from the Sun's rays

▼ *These cars are ready for the start of the Darwin to Adelaide race in Australia.*

solar panels

Faster and faster!

In the Australian race in 2003, Nuna II raced even faster than before. How did it go faster the second time? Three important changes were made. The labels below show what these were.

Nuna II fact file

First raced: 2001

Best average speed (2003):
97 kilometres (60 miles) per hour

Energy source: sunlight

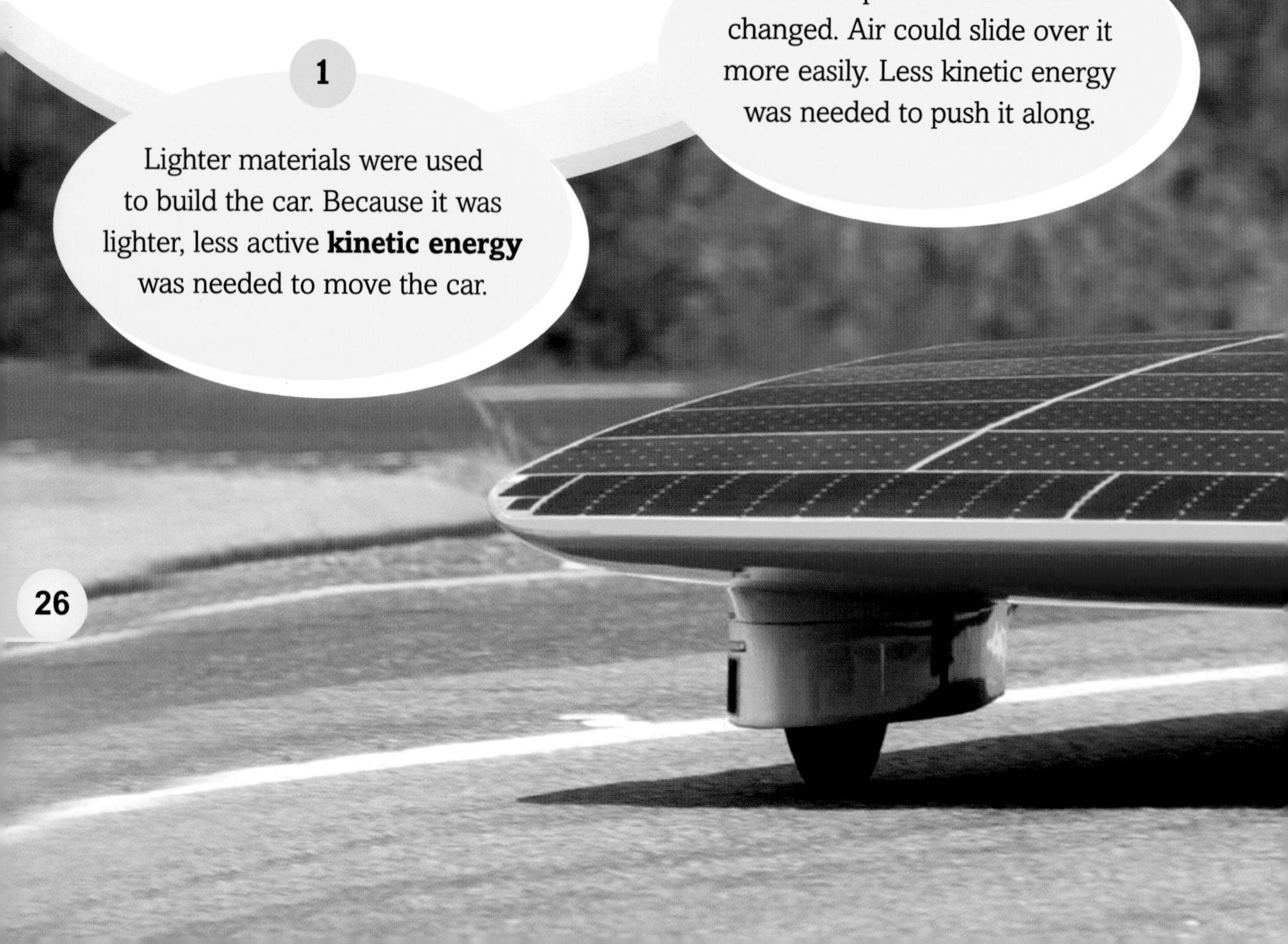

2

The shape of the car was changed. Air could slide over it more easily. Less kinetic energy was needed to push it along.

1

Lighter materials were used to build the car. Because it was lighter, less active **kinetic energy** was needed to move the car.

It would be difficult to use Nuna II every day. There is only room for the driver. There is no space for shopping bags. It is too wide and long to park easily.

Of course, the cars we drive today also looked weird when they first appeared. Who knows which of today's wacky machines will be a common sight in the future?

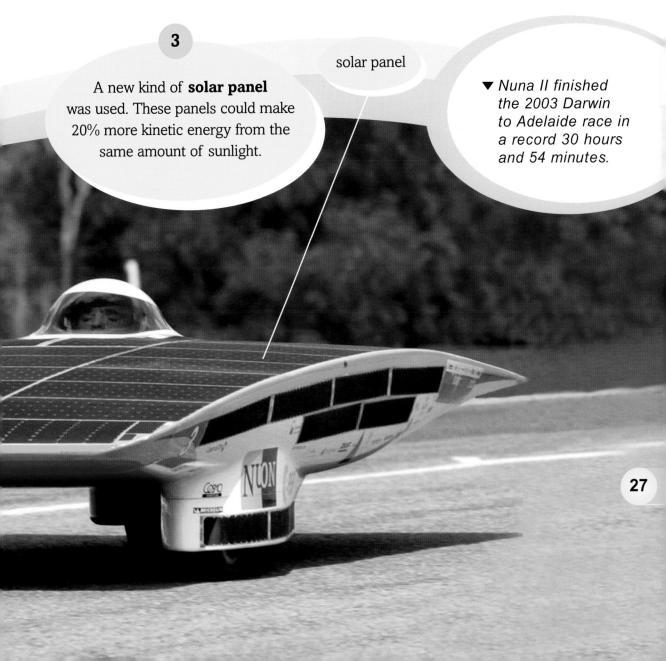

3

A new kind of **solar panel** was used. These panels could make 20% more kinetic energy from the same amount of sunlight.

solar panel

▼ Nuna II finished the 2003 Darwin to Adelaide race in a record 30 hours and 54 minutes.

History of wacky machines

Humans have used **energy** in many different ways through the years. This timeline shows some of the wacky machines that have been invented in the last 250 years.

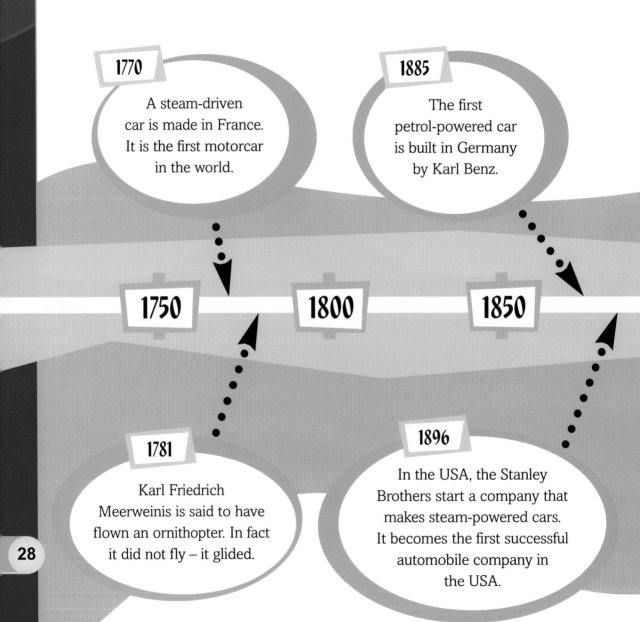

1770

A steam-driven car is made in France. It is the first motorcar in the world.

1885

The first petrol-powered car is built in Germany by Karl Benz.

1750

1800

1850

1781

Karl Friedrich Meerweinis is said to have flown an ornithopter. In fact it did not fly – it glided.

1896

In the USA, the Stanley Brothers start a company that makes steam-powered cars. It becomes the first successful automobile company in the USA.

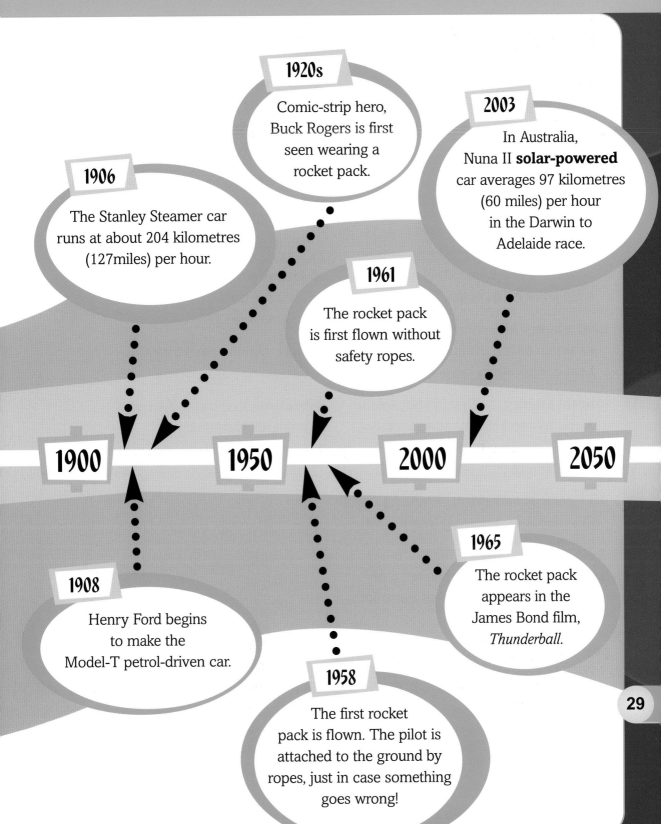

1920s
Comic-strip hero, Buck Rogers is first seen wearing a rocket pack.

2003
In Australia, Nuna II **solar-powered** car averages 97 kilometres (60 miles) per hour in the Darwin to Adelaide race.

1906
The Stanley Steamer car runs at about 204 kilometres (127miles) per hour.

1961
The rocket pack is first flown without safety ropes.

1900

1950

2000

2050

1908
Henry Ford begins to make the Model-T petrol-driven car.

1965
The rocket pack appears in the James Bond film, *Thunderball*.

1958
The first rocket pack is flown. The pilot is attached to the ground by ropes, just in case something goes wrong!

29

Glossary

energy ability to move something or make something happen. You get your energy from food.

friction when two things rub together. The force of friction will slow two moving objects down if they are touching each other.

fuel something that produces heat or power when it burns. Fuel is used to keep a fire or an engine going.

heat energy the energy in heat, such as steam or fire. Humans also lose heat energy when we get hot.

kinetic energy energy of movement. A rolling ball has kinetic energy. So does a human running along, or a river flowing.

lift upwards movement. Anything that flies needs lift to get up into the air and stay there.

light energy energy in the form of light. Light energy reaches Earth from the Sun.

potential energy stored, or possible, energy. When stored potential energy is used, it becomes kinetic energy.

solar panel special panel that traps the Sun's rays.

solar power power from the Sun's rays. Energy from sunlight can be used to heat homes, water, and power engines.

sound energy energy in the form of noise or sounds. Sound travels through the air as vibrations. These vibrations are a form of energy.

thermal energy energy of heat. Cooking on a barbecue uses thermal energy.

thrust sudden push forwards, backwards, or upwards.

vibration fast backwards and forwards movement.

Want to know more?

You can find out lots more about the wacky side of science:

Books to read

- *Heat and Energy*, by B.J. Knapp (Atlantic Europe Publishing Company Ltd, 2001)

- *Horrible Science: Fearsome Fight for Flight*, by Nick Arnold (Scholastic Hippo, 2004). The fun side of human attempts to fly, with as many crashes as take-offs!

- *Horrible Science: Shocking Electricity*, by Nick Arnold (Scholastic Hippo, 2000)

Websites

- www.flying-contraptions.com
  Find out about the "rocket belt", as well as some other weird machines.

- www.stanleymuseum.org
  The website for the Stanley Museum.

- www.ornithopter.com
  To learn more about ornithopters.

Learn how the world's top sportspeople – and you and I – convert food into energy for movement and growth in *Training For The Top*.

Light is a form of energy. Find out how it travels from a star to your eye in *Voyage of a Light Beam*.

Index